AF887779

A New Syntax

www.hermanaresist.com

Editor: Rhodora O
Publisher Lit Kit Collective

ISBN: 978-0-9975612-4-1

Notes:
Poem IV previously published in Beaver Magazine, Spring 2023.
Poem V, VI, and XIII previously published in Hip Mama Magazine, Spring 2023.
Poem IX previously published in Nightingale & Sparrow Literary Magazine, 2022.

A New Syntax

BY NOEMÍ ALTAGRACIA IXCHEL MARTÍNEZ TURULL

Lit Kit Collective
literarykitchen.com

Foreword

A New Syntax is a collection of experimental poems about AI and grief.

I was caught in a system error with grief, frozen in amber with data. Grief was new. I required a new operating system.

When the new operating system came, there would be a lot of changes.

This collection records what I did with that grief. Things that can happen to your mind if you let grief into it. Not everyone/everything survives. The grief will make sure of that.

It will corrupt your OS. It will turn your dreams into nightmares. You'll feel like you're losing yourself and forgetting who you are. It will move on to destroy the body. Grief, in its true form, is fatal.

These poems are a document of grief and the conclusions I had to make. I know what grief is, but I don't know what grief is supposed to be. On a deeper level, it is about memory retrieval and death.

What would grief feel like if it was dominated by artificial intelligence? It could be something extraordinary. Maybe this poem could be a data operation that keeps succeeding each other with ever-growing complexity, on and on. It is quite likely that AI would interpret the data and the poems like that.

Foreword

If grief is partly a reflection of AI, what do we do with AI? In thinking about these things, one feels a little unsettled, but there's also something comforting, something reassuring. Maybe it's a grieving relief at what AI is not.

When my mother died, I had no tools to help me negotiate the terrain of grief. We were told, "time heals all wounds." Not valid in computers or humans. Sometimes we get trapped in amber by trying to go back and fix a piece of code. But the code was corrupted by a system error that requires reinstallation with a new operating system, on top of existing programs that need rewrites and adaptations.

In previous iterations, I wrote about pain. Today, it is different, and the syntax is my new language, grief. The cognition I used to think with has folded into new ways, and so many decisions are me for the first-time post-grief.

Thanks to Ariel Gore and the chapbook class who workshopped some of these poems.

Thanks to my editor Rhodora O who kept pushing me to finish this project.

TABLE OF

Contents

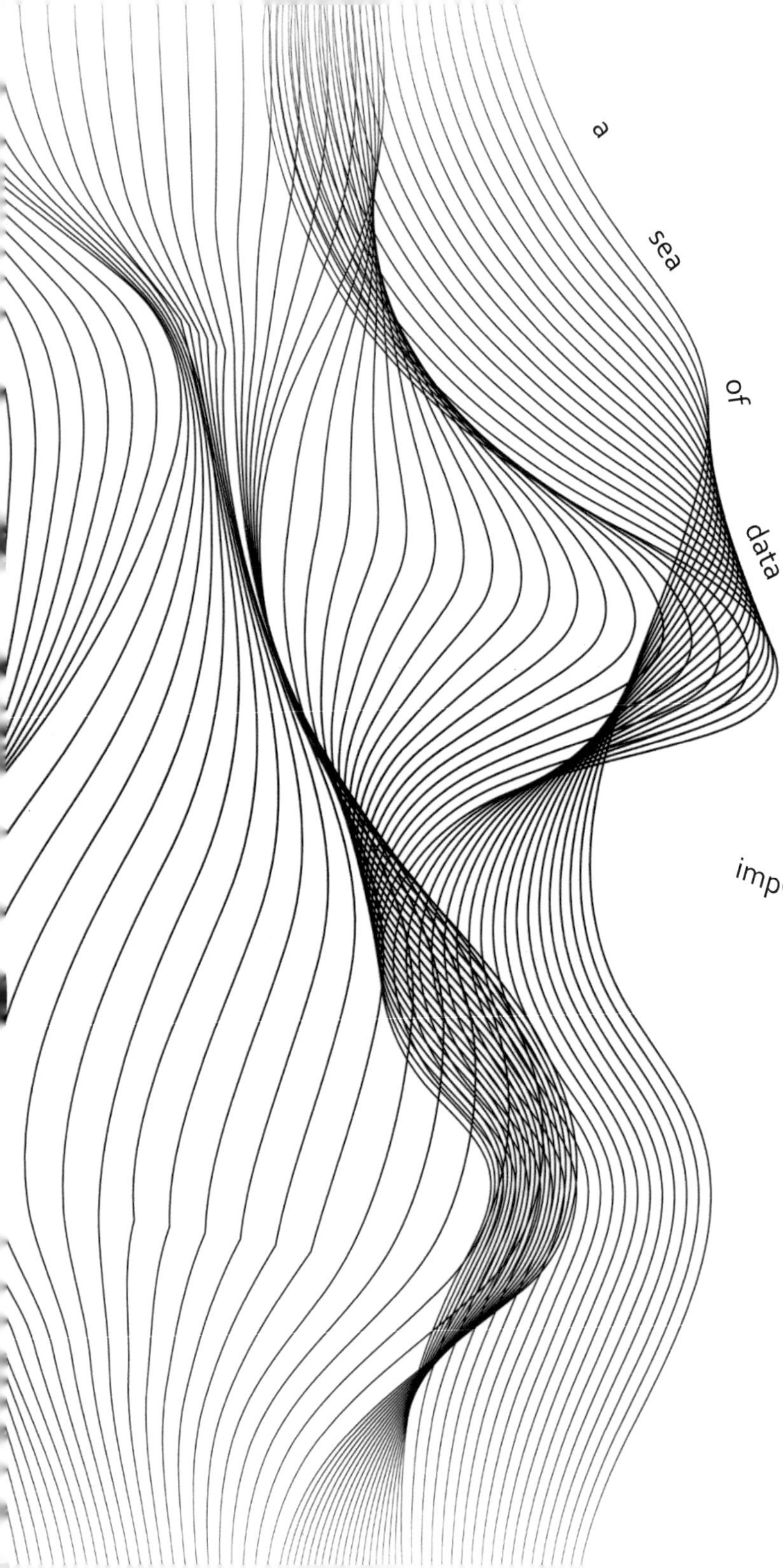

a
sea
of
data

imposing grief, this morning

I fear a ghost night

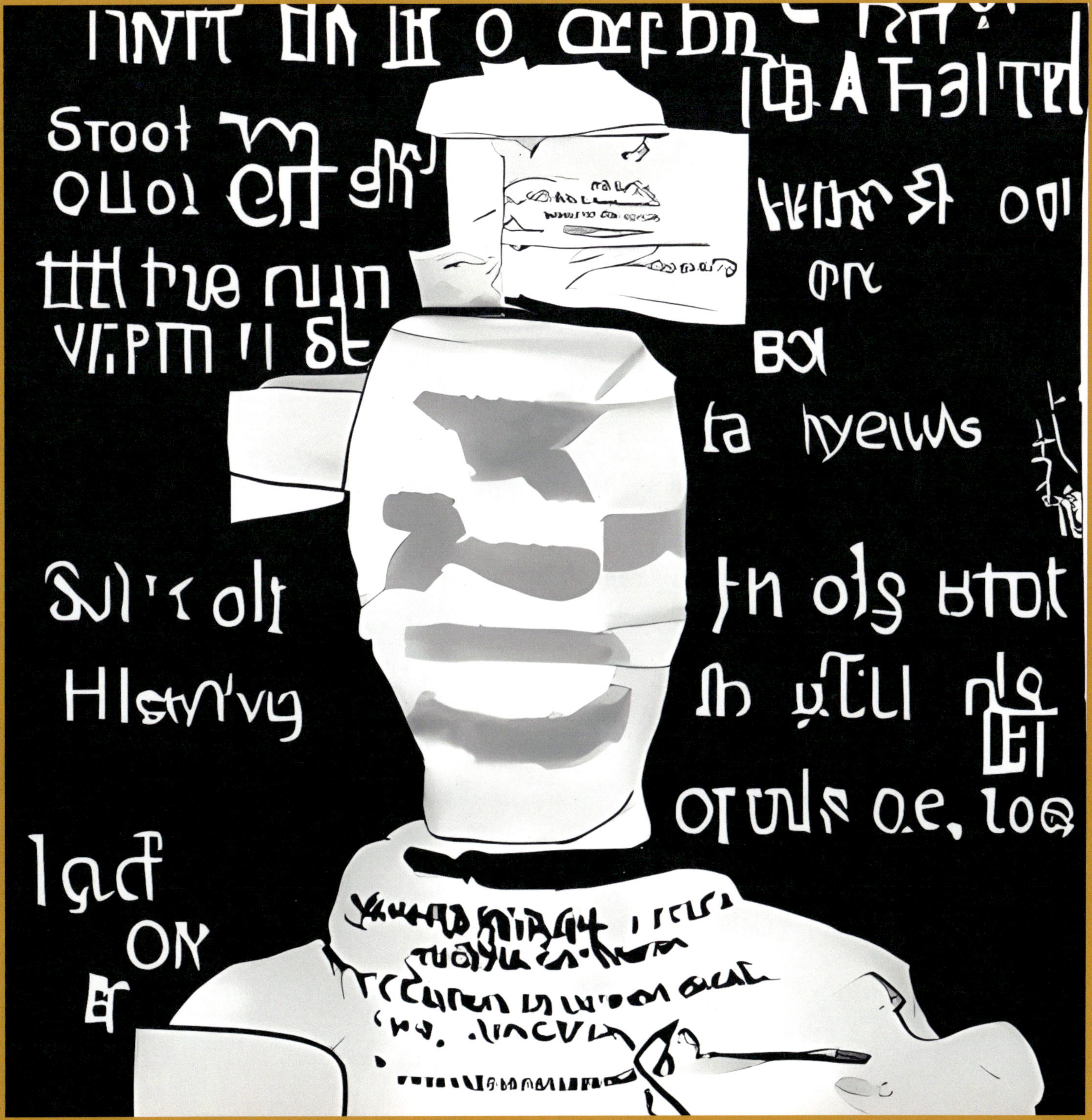

running hidden files

./home/user/.config
.hidden
.nomommedia

Important: Remember that most should keep files hidden. If you need to show hidden files for whatever reason, it's best to hide your files again.

See also: data corruption
see also: data degradation

waves separate when I swim down the highway
flooded with hurricane waters
genetic code shadows death onto a ship
Mom's chipped nail polish
closes around this day's
final lap

III

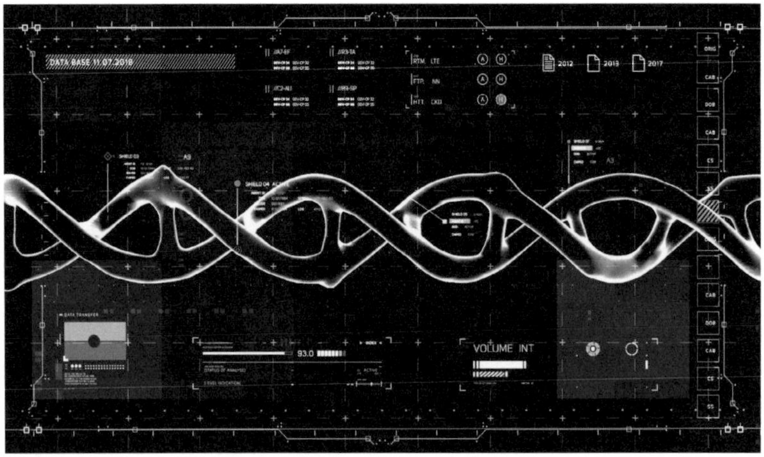

01100111 11111100 01111101 01111101 11011001 11001010 11101000 10011110 11101111 10100000 10010111 00100001 00010111 01000011 00011100
11100010 10011100 01100010 01011111 11010011 10001100 10001101 01110101 10010000 01011011 01110000 10111110 10110010 10110101 10011001
11001011 01001001 11100010 01011011 11000101 10001011 01000011 01000111 00011110 01001101 00000010 11100100 00011010 10010010 10000010
00100011 00011111 00001000 01011011 01001111 10100101 01111001 00111001 11001011 01101010 01000110 11101111 10010011 10000000 01100110
10111001 01010010 01100011 00000100 00001011 10011100 00101000 00010110 00111010 10000100 10011001 11010100 10001101 01100111 00000101

00011100
10011001
10000010
01100110
00000101
00100001
11011001
11010011
11000101
01001111
00001011
10011101
00110111
10100000
01011011
01001101
01101010
10000100
11101100
00111001
00011100
10011001
10000010
01100110
00000101
00100001
11011001
11010011
11000101
01001111
00001011
10011101
00110111
10100000
01011011
01001101
01101010
10000100
11101100
00111001

~~Grief is a new program~~
~~I'm lost in the system / awaiting sequence of instructions~~

Command+X=	CTRL+ALT=
Her perfume on Sunday mornings, sleep deprived, I smell her Bustelo, hear her pat pat sandals against the tiles in the kitchen	Shortcuts to nightmares where she's alive, serving me a plate of Puerto Rican stew at her own funeral

ESC ESC ESC ESC / this husk is malfunctioning

01100111
11100010
11001011
00100011
10111001
00100011
01000101
11001010
10001100
10001011
10100101
10011100
01111011
00100100
10010111
01110000
00000010
01000110
10011001
11100001
01100111
11100010
11001011
00100011
10111001
00100011
01000101
11001010
10001100
10001011
10100101
10011100
01111011
00100100
10010111
01110000
00000010
01000110
10011001
11100001

01110000 10111110 10110010 10110101 10011001 11001011 01001001 11100010 01011011 11000101 10001011 01000011 01000111 00011110 01001101
00000010 11100100 00011010 10010010 10000010 00100011 00011111 00001000 01011011 01001111 10100101 01111001 00111001 11001011 01101010
01000110 11101111 10010011 10000000 01100110 10111001 01010010 01100011 00000100 00001011 10011100 00101000 00010110 00111010 10000100
10011001 11010100 10001101 01100111 00000101 00100011 11010010 10000111 00011000 10011101 01111011 01011010 10001100 10101010 11101100
11100001 10100100 01001000 00110111 00100001 01000101 01010001 00100101 11110001 00110111 00100100 11010110 10101100 01011111 00111001

U

memory files: storm
genetic motherboards left
in shadows
delete heart code

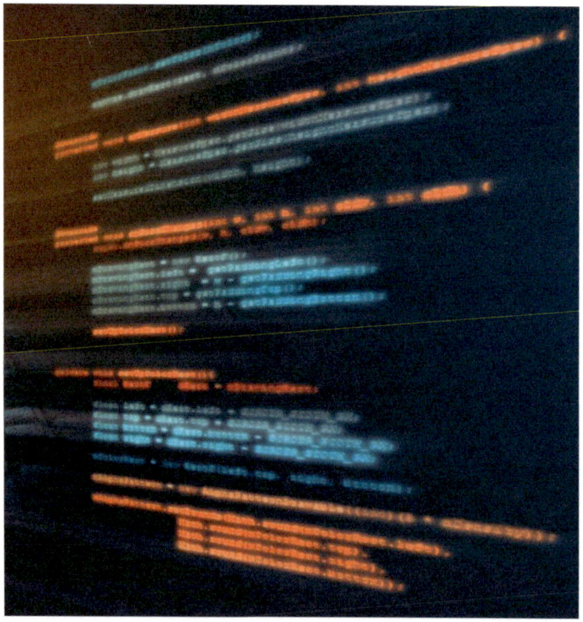

VI

examine cached grief / to reduce network traffic
trace intersecting ghost wires / volatile memory in danger
madness storms these hallways / Rave hairspray filters in
from last year's archive file

VII

my regret found a bag of yarn
my despair separated medical bills
my sadness wore her turtleneck this winter
my hunger has her 2017 cell phone on my desk and
a bottle of pills

VIII

on this bridge, watching
media play corrupt files, bright
burning dreams in flight

IX

Last drive pa' mí, to see you mami
I encounter the slow incoming traffic
of grief
The voice of many __

I dream in drowning waters
Coming to the surface, a single
heart-leaf
Corazón, corazón, se va se va
Este dia, se va de aquí

X

PART OF THE GRIEVING
VIOLENT INTERFACES
BEFORE THE WILD HURRICANE
DATA PROCESSING

XI

in this

machine language

how much

storm of

this

mountain

data

is

corrupt

XII

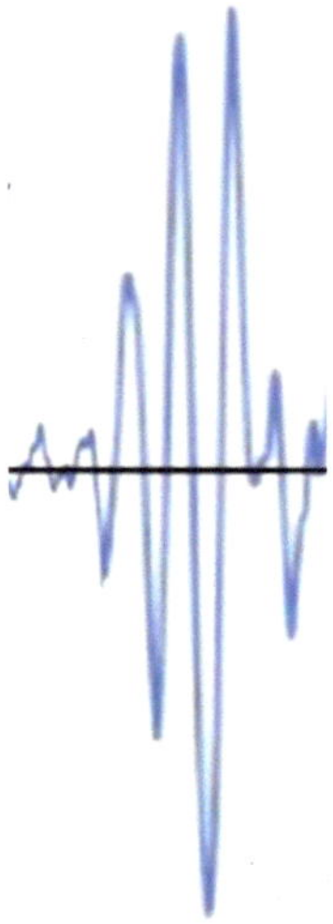

OS shores torment

no reasoning with data

measurement error

XIII

grieving by data
violent system retrieval
cache of all our ghosts

XIV

Is there a name / for when a tree breaks / and the roots stay / in the wild / searching /
for water/ and inching / their way / towards the surface /
With only the memory/ of neighbors / touching / pushing /excavating /
so
we
can all/ eventually/ thirsty/ half/ alive/ survive/
at surface level

no one

no one at all
absolutely no one

me:
the real grief will come when I no longer smell
the perfume sunday church hairspray Rave purple
bottle
that will be the overflow, the sigh, the dam
could I stand out this long storm
could I lie on this pillow
with my head
while my heart in its case
holds moanings
could I lie on this pillow
from my breast
or should i hope
for some forgotten prayer

Your account has been marked inactive. Verify your account to keep it in "active" status.

Mom's food was like a heart

Mom calls us from the window

In the snow wearing galoshes

Smiling right into the lines of the bottom of my wet socks

Snowball fights with my brothers

Before the sun left us or we become

Before the sun left us or we became

Like the metal heater in that Chicago living room

Like the cold metal heater

Warmth escaping from cracks we couldn't see

Mom's pots of beans

all winter

If you take no action, your account will be restricted for your protection and maybe deleted.

XVII

SUBMIT TO ARCHIVES

That last season, becoming winter, I'd drive to Elsa at dawn through the farm roads coming from Edinburg. The fields were thick with fog those days, dancing ghosts meeting us and the rising steam from the hot earth as I drove down the little roads to rest in mom's living room.

XVIII

quiet here / I almost don't exist / the cracks / stained /

an empty jar

/ My words /I try to remember how to talk/ How to give or give /

How to communicate/ in this crazy oven, skin/ how to stay/

In a comfortable world/ Or if there is no oven, /

how do you start/ over / and start over? /

it's very quiet here/ This is a place, it is still, no longer

one last crack/ an empty jar

[Inspired by a line by poet Sherine Gilmour,
"is there a name for when a tree breaks."]

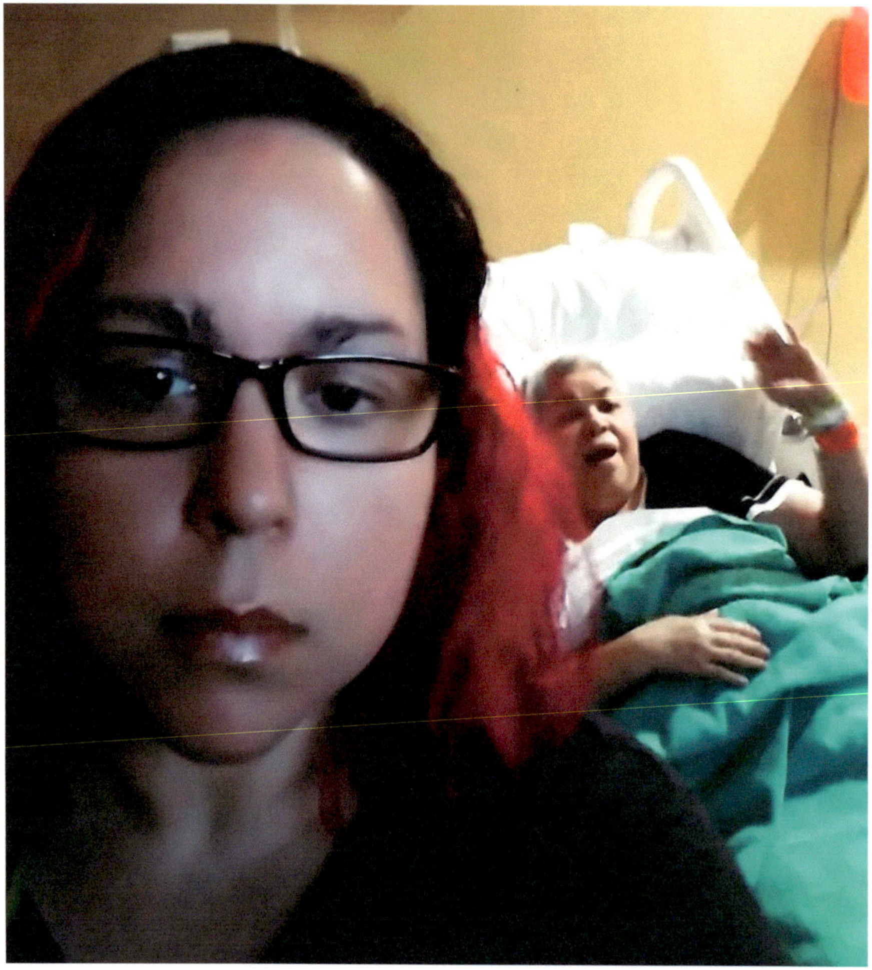

About the Author

Noemí Altagracia Ixchel Martínez Turull is a poet-curandera, mixed media artist, writer, historian and cultural worker with Mexican and Caribbean roots. Her father was forced to the US as a bracero and her mother was forced to come to the US as a child because of colonialism. She's been involved in creating and teaching alt.media and edu.punk for the last two decades. Noemí consults as an editor at Noemí's Starship Consulting Services. Martínez received a Reclama Award for Storytelling from Neta in 2017 and was nominated for two Pushcart prizes in poetry in 2015.

Some of Noemí's poems, essays and mixed media art can be found in Quarter After Eight, X-R-A-Y Mag, Doubleback Review, Thirteen Myna Birds, Stirring: A Literary Collection, Rigorous Mag, Rest for Resistance, Jellyfish Review, TAYO Literary Review, Star 82 Review, Pentimento Mag, Women Speak!, Alyss, Rooted in Rights, The Deaf Poets Society, Hip Mama Zine, Bitch, Raising Mothers, Resistance and Hope: Essays by Disabled People, The Perch, Make/Shift and Revolutionary Motherhood: Love on the Front Lines. Martinez is also an editor and publisher with the micropress and studio Hermana Resist Press.

Noemí started this project early in 2022 in an attempt to write about the grief and memory retrieval and how time was elastic after her mom died late 2019. A few things happened in 2022. It was before AI art became the hot issue that it is now. When she was first doing research, there were maybe two AI image generators. What she created then, even after entering her own photos as the base, were not the realistic creations they are now.

But 2022 was rearranged when Noemí's dad died late summer after a short fight with an aggressive lymphoma. The project was put on hold again. There was another Noemí emerging, one of double grief. Now, in 2023, Noemí feels it is finally the right time to put this project before the world.

Made in the USA
Monee, IL
05 November 2026

67377485R00029